Indian
Cartography

Indian
Cartography

Deborah A. Miranda

Greenfield Review Press

State of the Arts
NYSCA

Publication of this book has been made possible, in part, through a grant from the literature Program of the New York State Council on the Arts.

© 1999 Deborah A. Miranda

Published by The Greenfield Review Press, P.O. Box 308, Greenfield Center, New York 12833

All rights reserved. No part of this publication may be reproduced or transmitted in any form or by any means, electronic or mechanical, including photocopying, recording, or by any information storage and retrieval system, without permission in writing from the publisher. Reviewers may quote brief passages.

Publication Acknowledgments

Some of the poems in this collection have appeared, sometimes in slightly different form, in the following publications:

Bricolage, Callaloo; Calyx; Cimarron Review; Durable Breath, edited by John Smelcer and D.L. Birchfield (Salmon Run Press, 1994); *News From Native California; Weber Studies Journal*.

"Baskets" was also published as a limited-edition letterpress broadside by May Day Press.

ISBN 0-912678-99-2

Cover and interior design by Sans Serif, Inc., Saline, MI

for my tribe,
my family,
our children.

Contents

Acknowledgments *ix*
Introduction *xi*
Author's Note *xv*

Certain Scars
Looking For a Cure *3*
Stories I Tell My Daughter *5*
Prayer For the Fourth of July *7*
While You Were in San Quentin *8*
Strawberries *11*
Sea-Tac Airport, May 1974 *13*
Hunger *16*
What Part of Me *20*
Wildflowers *22*
Formula *25*
Venom *29*
Certain Scars *32*
Lost Language *34*
Finishing What He Started *36*

Bodies of Water
Sometimes the Open Hand of Desire *41*
Refuge *42*
Summer Solstice *43*
Bodies of Water *44*
Shame *45*
Commencement Bay *47*
The Territory of Love *48*
A Walk *49*

Grief *50*
Sorrow As A Woman *52*
A Ceremony for Crying *54*
Heartwood *58*
Vernal Equinox *60*
Going On *62*
The Night of No Shadows *64*
Winter Solstice *65*
Three Poems for April *66*

Indian Cartography

Naming the Nameless *71*
I Am Not A Witness *73*
Without History *74*
Indian Cartography *76*
Migration *78*
January Cusp *79*
Correspondence *80*
After Colonization *82*
ghazal 8/7/94 *83*
For My Other Grandmother *84*
Gone Dancing *86*
Riding the Back of the Universe *88*
What Is Possible *89*
Tehachapi Night *90*
Baskets *92*
I Dreamt Your True Name *94*
Burial Ground *96*
Waking *98*

Biography *100*

Author's Note

A few years ago, I sat at lunch with a group of writer-friends, one of whom is also California Indian. When we spotted another Native woman walking by, we invited her to sit with us, and in the course of the usual introductions, a very familiar dialogue repeated itself: "What tribe are you?" "Esselen—(anticipating, and getting, the usual blank look)—that's a California tribe." "California! I heard all those Indians died!" My friend and I looked at one another, and, smiling, announced, "Well, here we are!"

It was an affirmation that, on the surface, seemed obvious enough, and we gave it cheerfully. But sometimes I think that when California Indians say those words, something in the earth stirs, and the air becomes electric with existence. Because the truth of the matter is, for many tribes in California the words "all those Indians died" are horribly real. Our lives, and our ability to assert our identities as tribal people, remind me of ingenious, stubborn plants cracking through the thick pavement of history. We disturb the facade of conquest—challenge the myth of extinction—but though we still exist, we are not undamaged or unchanged. The Indians in California were never meant to survive, much less experience a resurgence of culture and literature heading into the 21st Century.[1] Our demise was legally and officially planned, executed, and very nearly carried out.

Every tribe in the Americas has its own account of colonization; the details vary with geography or chronology, but what never changes is the result. I want to tell the story of things intrinsically evil and hateful that happened to us. Twin disasters. In Northern California, one evil is called the Gold Rush. For my people, the Es-

---

[1] For beautiful proof of this blossoming, see *The Sound of Rattles and Clappers: A Collection of New California Indian Writing*, edited by Greg Sarris.

selen, Ohlone, Costanoan and Chumash, the name of evil is Missionization.

The incredible mythology that surrounds the Missions of California has been perpetuated by the same institutional fantasies that insist colonization is not a crime, but an "evolution" of history and human civilization. To this day, schoolchildren in many parts of California are required to create dioramas of Missions, complete with happy, productive Indians working in the fields, at looms, and worshipping in the sanctuary of a benevolent Catholic god. What the children do not portray, and are not taught, are the scenes of flogging, rape, kidnapping, European-carried diseases like syphilis and smallpox, torture, enslavement and starvation which were also parts of the Missionized Indian's day. Over one million Indians lived in California in 1700, before the first Spanish settlement. Between 1700 and 1900, the indigenous people in California passed through the hands of Spanish, Mexican and American governments, each one more brutal than the last (the door-to-door murdering of Indians during the Gold Rush occurred under American rule). Secularization, which formally ended the Mission era in a series of legal acts between 1834–1836, saw only a handful of survivors left to accept their "emancipation".[2]

My relatives were some of 20,000 Indians who walked out of this evil alive.

I am a mixed-blood woman, the daughter of an Indian father of Esselen/Chumash ancestry, a white mother of European/Jewish ancestry. I know that ancestors from both sides of my family move inside me, give me their best and worst traits. Perhaps because I look like my father, am seen by others as Indian, I identify most strongly as an Indian woman. But this identity is where my heart and blood live, too. My life has been an embodiment of separation,

---

[2]Information derived from *Ethnology of the Alta California Indians (Spanish Borderlands Series)*, ed. by Lowell Bean and Sylvia Brakke Vane. I'm grateful to Alan Levanthal of San Jose State University for directing me to this updated information.

division, reconciliation, and loss. I am Indian: I am mixed-blood: I am Indian.

My grandfather, Thomas Miranda, came from the Esselen Nation, around the Monterey/Big Sur/Carmel River area. This part of my family was "taken into" the Carmel Mission. My grandmother, Marquesa "Keta" Robles Miranda, came from the Chumash of the Santa Barbara/Santa Ynez River area further south; these people "belonged" to the Santa Barbara Mission. Growing up, this was all I knew about my Indian ancestors for a very long time. I was told, "Yes, you're Indian, but they're all gone now, so it doesn't matter."

But it mattered to me. Separated from my Indian father after my parents divorced, I began asking my mother questions about him, and she managed to contact him and bring him back into my life. At thirteen years of age, I finally met someone who looked like me. It was the beginning of a frustrating search for my people. For years I was told the Esselen (also known as Costanoan) had been terminated as a tribe by the United States Government, and the Chumash, my grandmother's tribe, were understandably leery of registering me without very clear proof of relationship.[3] I was experiencing the legacies of colonization, some of which were peculiar to Mission Indians. Families, supplied with Spanish names, language and the Catholic religion in the missions, "passed" as "Mexicans" in order to survive the violent, deadly anti-Indian racism in California; "passing" meant recording your race as Mexican on all official records, the very records the U.S. Government would later demand as "proof" of Indian ancestry. In large part because the few remaining tribes with land were constantly shorted and cheated by the U. S. Government, Indians have been forced to demand "blood

---

[3]Confusion about tribal names is not uncommon when discussing California Indians, especially tribal groups along the coast, which often had fluid kinship and land boundaries. The fact that the Esselens had been declared extinct made it impossible for my grandfather to register as such; however, the Ohlone-Coastanoan Esselen Nation is our official title, indicating a close relationship which is more inclusive than exclusive.

quantum" certification from each other, fostering mistrust and bitterness. The worst legacy of all for California Indians whose ancestors emerged from the Missions was the basic loss of familial connections through a diasporic, desperate scattering of tribes without a landbase.[4]

All these circumstances hampered my efforts to find the people, and the places, whose absence left me incomplete. But between my father's recollections of events, my mother's "bull-headed" determination, and a fortuitous encounter with Lorraine Escobar, tribal genealogist for the Esselen Nation, an amazing thing occurred: Lorraine and my mother found, resting in the San Bruno Archives, one sheet of paper, the document which showed that my grandfather had registered himself and his sons as members of the Costanoan people—with the appropriate "legal" proof. Lorraine and the tribe welcomed us with the words, "We've been looking for you all along." The Mirandas, it turned out, were one of the few remaining major families of the Esselen Nation. They would have taken us in without that piece of paper. But the documentation mattered to me: it meant my grandfather had known who he was.

Recently, my father has told me stories about his Chumash mother; how she and her mother would supplement insufficient groceries with the acorn harvest, cactus apples, pinon nuts, salmon my father and his brothers literally picked up out of spawning streams. Marquesa knew who she was, too. She ground and prepared acorns for her sons, and they ate the resulting thick soup or

---

[4]The Chumash, once the most populous and successful tribe in California, have one small reservation, on which about 150 tribal members still live. In 1988, Bruce Miller estimated there were about 1500 people of Chumash ancestry living in the San Luis Obispo, Santa Barbara, and Ventura counties. Other researchers cite much higher figures of mixedbloods living in the area. Again, when dealing with Indian populations, historical and political influences make more precise numbers very difficult to decipher. The Esselen Nation was declared extinct long ago, and has no land at all. About 300 tribal members are currently petitioning for Federal Recognition.

mush the same way their ancestors had: a skillful scoop with two fingers, not a drop wasted. She spoke Indian with her mother, and though my father remembers only one or two words, I hear the inflections of his mother's tongue each time he speaks; a soft, slightly slurred, familiar accent.

My grandmother's knowledge of native language and foods, my grandfather's assertion of his Indianness, are nothing less than miracles. I wish there were some way to express that word, "miracle," in all its unimaginable and unexpected hope! Sometimes I think I can begin to grasp the terror and cruelty that existed in the Missions— I can at least name it, the truth is being told now—but what I can't understand is how any of us survived such a system. Nothing about Missionization encouraged an Indian identity; we were denied, destroyed.

Or were we?

Because some of my relatives survived the Missions, survived secularization, survived the poverty, prejudice, alcoholism, diabetes, and abuse that followed and still persists, I am here. Because our tribe has begun the process of applying for Federal Recognition, we have found relatives, stories, and strengths that we didn't know we had. Because the color of my skin, my eyes, my hair, called out for those who knew me, because my longing for tribal connection ached in my bones, because of some spiraled, resilient chain of events that led me home, I know who I am: a mixed-blood woman with passionate ties to her Native ancestors. And I want these poems to say those words that testify to a miracle, that make song out of quivering air: *Here we are, here we are, here we are.*

Deborah A. Miranda
May, 1998
Tacoma, Washington

Introduction

Deborah A. Miranda's *Indian Cartography* is an impressive debut. She is among the most resonant of the younger generation of mixed-blood voices from native North America. Not since Wendy Rose has California offered such an original and invigorating stylist. She seems most at ease working within the prose poem structure, but occasionally risks attempting the lyrical frame. She takes her metaphors seriously and they emerge at the point where they are most needed. Nevertheless, just as names and places are the definitive characteristics in the literature from the ancient stories and poems to those of the present, we discover that names and places give shape and character, range and depth to her work.

Thus the poems grow with the season and ripen like fruit from the trees found in the land of her Indian ancestry, the Esselen Nation (also known as the Ohlone/Costanoan). Her grandfather was from the Monterey Bay area and her grandmother from the Santa Barbara/Santa Ynez area. These are her father's parents. Her European inheritance stems from her mother, who is French and Jewish. I only mention this mixed blessing because it appears to be a central force in what conflicts feed her imagination and add substance, direction, and focus to her narratives of these two poles that sustain her world view. When her family traveled north from California to Washington when she was a girl, the names and places there inspired her aesthetic impulses and extended her frame of reference and path.

I believe what first drew me into the world of Miranda's poems was the surprise of the special place it took me to, which is, no doubt, what attracts us to any art form. As the spirit of art would have it, that special place turns out to be the mundane world we all live and dream and work in. For art is as close or far away as a child tightly holding a toy while asleep in a crib, to a wife beside herself at losing an opportunity to turn words into a healing ceremony, to the lost chance to say, "I love you too much," to a dying loved one who has spun havoc for nights through one's dreams.

However, since poetry arises from below the surface of language in the rhythms of the oral tradition found in word, line and stanza, presentation and process are great sources of energy and innovation. The way the poet chooses to arrange the words so they will dance off the page like a family into our mind's ear and eye and hand also shows how much the art is a spiritual quest. This is, of course, the case regardless of how tense the situation that triggers the essence of what is to be perceived. The process that brings us closer and closer to the place of original utterance captivates us. So it is that words in combination and their unified dance most suggest to us the experience's sense of reality, its body and soul in spontaneous motion and rest, its details of the specific image rooted in the physical. But in addition, it is also our gut-level response that helps open the door to the images and feelings and the pattern of their rhythms. This is why we can say the poem is like a gene-pool sonata or writing transformed from a cave wall. To me this is essentially what Ezra Pound meant when he said that the poets—to really connect with their audience—must "make it new." He was implying that you can only make something new if you uncover the words from under some hidden rock where the ancient voices live. Looking at this from another perspective, one can say Miranda has dug deeply and tirelessly into her own inner nature to reach the primal sources for her songs and has lifted that rock to see what lies underneath. She has recognized intuitively that nothing will be found within unless it can be connected to what is outside, something the ego-maniacs never seem to learn. Therefore, she is successful in her efforts to probe the darker recesses of self and world. But it is time to let her speak directly to you about what she attempts with her art made of words. Now to quote a few lines where the evidence of these claims is clear.

*

"... She is always just on the other
side of knowing, separated by a wall
of flesh or maybe layers of bone,
marrow thick at the center.
Bleeding is too small a word."

*

"... The territory of love does not take kindly to being
colonized. To live there, you must give what you think you
have learned, take back what you fear is forsaken: understand
you can only win by losing."

*

"In my father's dreams
after the solace of a six-pack,
he follows a longing, a deepness.
When he comes to the valley
drowned by a displaced river
he swims out, floats on his face
with eyes open, looks down into lands not drawn
on any map. Maybe he sees shadows
of a people who are fluid,
fluent in dark water, bodies
long and glinting with sharp-edged jewelry,
on the stories of our home."

—Duane Niatum
Marysville, Washington
Moon of Budding Trees, 1998

Personal Acknowledgments

I would like to thank the many people whose encouragement, expertise and tenderness have helped guide me: Joseph Bruchac, Lorraine Escobar, my Esselen Nation family, Mei Mei Evans, my companions at "Flight of the Mind," Malcolm Griffith, Janice Gould, Juan Guerra, Ruth Gundle and Judith Barrington, Amelia Haller, Geary Hobson, the Jansen Clan, Bret Keeling, Catherine Michaelis and Kim Newall, Tiffany Midge, Madgel Miranda and Alfred Miranda Sr., Pat and Sally Miranda, Tom and Rose Miranda, the women of MUSE, Duane Niatum, the poets of Northwest Renaissance, Grace Paley, Carter Revard, Marjorie Rommel, Wendy Rose, my sisters and brothers Annette, Rose Marie, Louise, Lenora, Pat, Kacey and Al; Margo Solod, Robyn Train, and Marie-Elise Wheatwind. And always, thanks to my grandmother Doris Gano Yeoman, who passed away before this project was completed.

My deepest appreciation to the Native Writer's Circle of the Americas and the University of Oklahoma for the opportunity to publish, and for the warm reception in Norman.

Most of all, very special thanks to my husband, Daniel Miller, and our children, for loving and growing with me during the writing of these poems.

Certain Scars

Looking for a Cure

In this country that is and is not
hers, a woman sees a stream of bloody sparks
racing along an obsidian-slick channel.
Cut out of heat and desperation,
she can never map or anticipate
this path. She is always just on the other
side of knowing, separated by a wall
of flesh or maybe layers of bone,
marrow thick at the center.
Bleeding is too small a word.

Past hope of any tenderness,
the woman understands she exists
at the mercy of a rage
that threatens to carve up her soul
like cancer advancing in stages
of nameless pain. She dreams at night
of eagles with lost wings,
bears enslaved. There is a medicine
for this anger if only she could
remember it
or chant it
or locate the shady grove
where the plant still grows.

Don't tell her too much
has been forgotten,
there are no longer directions
to this place of resolution.
The woman dreams in daytime now
of a cave of sacred bones, walls
resonant with voices singing in a cliff.
Below, cold water rushes
through a green cut lush with potency.
She feels a path incised into her palm,
a course that flows directly
into the heart's homeland.

Stories I Tell My Daughter

1.

Once when I was eight we came home in the dark
up our winding dirt road through a tunnel
of thick spring leaves, and our headlights
turned maple and salal into a green
glowing vein leading us sleepily to our beds.

Suddenly the headlights caught an owl,
wingspan as wide as the windshield,
every creamy feather etched in perfection,
round eyes huge and yellow
with pupils black enough
to swallow me up. His curved beak
opened in surprise
at being face to face
with this steel creature whose own eyes
shown unblinking and wild.
                        The owl didn't cry out but I did,
lunging forward into the dash,
hands beating against what was invisible
between us. I shouted, *No!*
Don't—The owl spread every pinion,
drummed hard against the skin of air
between life and death,
lifted backwards—up—over

and we drove on home, giddy with fear.
All summer I lay awake on cool sheets, window open,
waiting for a low urgent call.

2.

At thirteen, I played the drums, first chair,
ahead of boys who called me squaw.
I stopped wearing braids,
stood with my back to them,
my hair thick down my back
like a cloak no looks could pierce.
Their dirty jokes and snickering
made my strokes tighter, sharper,
hands curved around the wooden sticks
in an easy grip, as if they were tools
I had always used.
                        Then there was heat
against my back, a low warning breaking
through the rhythm,
stench of singed human hair.
When I whirled around, it was Damon,
the one who always called to me in the halls,
I'll give you a little papoose, squaw!
holding a butane lighter, flame high,
orange and blue. In his hand
he held a chunk of black hair,
my hair,
burned off, smoking
in the still bandroom air.
                        That day
I took bloody sticks home to my mother,
who said she expected nothing less
from a girl
       who spoke
              to owls.

Prayer for the Fourth of July

With dawn the rain begins—softly at first, like tentative kisses beading up over the earth's dry skin. Overnight the temperature has dropped sharply. In the upstairs bedrooms of a house built from old pine, fir and cedar, a family sleeps deeply for the first time in a week.

A woman wakes to the dull roar of fans blowing down hallways, blowing on her children's chilled brown limbs. She gets out of bed, pulls crumpled blankets over the bare skin of a girl and a boy, turns the blade speed down. She hears, suddenly, the rain of her childhood. She smiles. She thinks how the coast is bathed in fog and mist, how wetness curves inland from Anacortes to Vancouver, green swath of a spoiled long weekend pushing all the way up against the Rockies. Tribes with names like Quileute, Skokomish, Nisqually and Puyallup sing in this weather.

Outside the window the woman sees dark faces under reed hats streaming with summer's richness, watching her. Now she remembers barbecues guttering in wet wind, hot dogs eaten uncooked in soggy white buns, her mother and step-father under some tree or in the car; enough of those fat brown bottles of beer to make it till dark when fireworks burst forth over the lake. She remembers flashes searing for unnamed moments, each raindrop plunging down, water into water, water into hair, mouth. All these Fourth of Julys are stored in a girl's cold body.

While You Were in San Quentin

Eight years for my father

1

I became a stepdaughter.
We moved to Washington State
in a year of drought.
I didn't have a brother and sister anymore.
I lost my front teeth.

2

One teacher said I was too dark,
too quiet, too slow.
Another teacher cherished me.
I learned to read.
We moved again.

3

Hannah became my best friend.
My mother had an affair.
This man molested me.
This same man molested my best friend.
She told.
Our mothers believed us;
the sheriff didn't.
We moved again.

4

You sent three cards.
In each one you called me
Darling Daughter.
I never wrote back.
Kittens were born under my bed.
We moved again.

5

I memorized tavern phone numbers:
Mecca, Ad Lib, Sugar Shack.
I grew my hair long.
My step-father left;
then, he tried to sell the trailer we lived in.
We bought macaroni, hot dogs, and ice cream
with food stamps.

6

My mother studied all night at the kitchen table.
She had a new boyfriend named Joe.
He was kind to us. I hated him,
his Oklahoma accent, the way
he wanted me to trust him.
In the summer we picked berries.
I told people my real father was Indian.
I told people I had six sisters,
named them.

7

A man walked on the moon again.
I failed the multiplication tables test over
and over.
I grew breasts, hips and got my period in fifth grade.
My grandmother bought me the ugly clothes
for fat girls.
I forgot what little Spanish I knew.

8

I took up drums instead of typing.
I waited for a new best friend.
I tried cigarettes.
I kept a journal of my dreams.
I began to wonder who I looked like.
I wondered if I looked
like you.

Strawberries

for H. L. S.

We wait apart from other kids
for the early bus to the fields;
it is foggy, and we both have sweaters
two sizes too small.
It doesn't matter.
We pick all day on our knees, me
on one side of the bushes,
you on the other.
Sun turns my bare arms cinnamon, burns
the soft Irish of your cheeks,
arms, tops of your ears. We take turns
guarding each other's half-filled flats
during that long walk to the outhouse,
keeping punchcards on strings
around our sweaty necks.

Sometimes you sing to me, verses
from vacation bible school about Jesus,
Michael, building an ark. We duck
low during rock fights, swear at yellowjackets
and small gray slugs. Once, two girls
are caught together in the cool brown
dirt between rows. You and I learn
a new word, the price of their kiss
whispered across hot fields.
We say it once, but not again.

Best friend: Mine, all mine.
It means we don't have to share
with anyone else. What do we have
anyway, except the weight of a full flat
divided by two, red stains
ground into our knees,
bright berries that break
like warm hearts
inside our mouths?

Sea-Tac Airport, May 1974

for Al Miranda, Jr.

I was thirteen years old, already as tall as I'd ever grow. I had a woman's body—small breasts, wide hips—and wasn't comfortable in it, hiding inside jeans, flannel shirts, baggy sweatshirts. You were three years old. Your body was compact and husky. You had my dark hair, my brown eyes, my strange olive skin, but darker. We recognized each other at the gate, though we had never met. Later, people would mistake you for my son.

While my mother chain-smoked off to one side, we waited all night on Concourse C for your plane from L.A. Was it a clear spring night? Was there rain? I paced the shiny floors in that unnatural airport light. I wonder if, while I waited, you slept on the plane in our father's lap, exhausted from a fight between our father and the woman who was your mother. Did he tear you from her arms, did you cry? Or were you secretly pulled from your bed in the dark, hushed into silence, carried out of the room you shared with your half-sister, who was not chosen, but left behind?

I wonder if you sat upright on the plane in your own seat by the window. Did you know how to cup your hands around your eyes and press your face against the glass? Did you peer out at the half moon, did you know those dark shapes floating below were clouds? At breakfast the next day, our first meal together, I showed you how to use a fork. I showed you how to peel off wet pajamas and underwear, change sheets without our father seeing. It was my first act of rebellion: protecting you.

That night in May you finally stepped out of the gateway, walking beside our father. I memorized your face. The way you held your body: your thick black hair, carefully combed to one side, wet and slick above a broad forehead, your large dark eyes that didn't seem to blink, your shoulders hunched. I shouldn't say I memorized you, little brother. I shouldn't say I recognized you. What happened was that you stepped out of the gate, and everything that defined your three-year-old soul flew across the white space between us, was engraved in the cells of my body. Did I love you then, in that instant? Was that when you mistook me for your mother, when I moved toward you? You put your hand right into mine, without hesitation: smaller, but the same color, same shape. Later that would change, and your hand would dwarf mine in size, become muscular from labor, dark from hot summers spent roofing, framing, from parties on beaches, fast transactions. I wonder, am I still written in the lines of your palm? What have your hands done that could erase me?

Our father introduced us as strangers streamed around us looking for friends, taxis, phones. *This is your sister*, he said to you. I knelt in front of you and watched your lips. My name in your mouth took on a foreign feeling, a distortion caused by the way you didn't know how to form sounds into meaning. How could you know how far you'd come, when you rarely left your apartment building? How could you talk to me when good behavior meant silence, compliance? I realized everything would be new to you: buttercups by the creek, cats, dirt not buried beneath cement or asphalt. You would preface each discovery with my name, demand labels and explanations. *Deby, look dis. Deby, tell dis. Deby, dis? Dis?* Later, after you learned to talk, you learned to lie. *No daddy, I didn't wet my bed. No, I didn't make a mess. No, it doesn't hurt when he hits me with the belt. No, I didn't touch the liquor cabinet. No, I didn't kill that frog, didn't stab it with a stick. No, I didn't lose my job—I quit. Naw, I only had to spend one night in jail, no sweat.* Still later, your mouth would your mouth would forget how to say my name. You would forget how to talk to me. Something else spills from your lips now.

We turned and walked away from the gate. Our father walked ahead with my mother, the woman you would claim as parent now to schools, on emergency forms, with friends. You and I walked behind, looking into dark empty restaurants, bars, gift shops behind locked grates. I didn't know yet that my heart had mistaken you for my son. I didn't know that in three years our father would take you from your bed, move away again. I held your hand all the way home in the backseat of the car. Later, in your new room, you wouldn't let go until you fell asleep.

You don't know him, but your nephew wants to do the same thing now; six years old, he grabs my hand in parkinglots, on hikes, at bedtime. When I go into his room in the middle of the night to pull up stray blankets or close a window, I find my son splayed on his back, black hair flung away from his wide, smooth forehead. One hand is closed around a toy truck. The other hand rests open, palm up, fingers gently curled around the space that my hand pulled away from a few hours ago.

Hunger

You came back for me one day:
drove up in a big white pick-up truck
with a new husband. I left them
behind in an instant—
Uncle Mike, Aunt Sandy,
the cousin my age who was so pretty,
so jealous. I climbed into that truck
and we drove away together.
I knew I belonged
only to you.

We began again,
in a small house
with my own bedroom
and a huge palm tree out front.
California spring.
Nests filled the branches:
robins, pigeons,
or maybe crows, I don't know;
I only remember
the constant cries
of hunger waking me each morning,
a rush of wings
as parents came and went
with food. I ran
on tiptoes
into the bright kitchen,
found your embrace,
and breakfast waiting for me.

I always wanted
yellow cornflakes in milk.
My appetite
came back, *how many bowls?*
you laughed, but you were pleased.
Afterwards while you cleaned

I went outside, dived
into the vacant
lot next door, reveling
in orange poppies, grass
tall as me. I tunneled, hid,
made caves of dry stalks alive
with grasshoppers,
potato bugs, spiders.
Much later, I emerged
flushed with secrecy,
anxious for lunch.
There was tomato soup
with crackers,
cold milk in a thick
glass tumbler. Your hair
was damp and fragrant
beneath a silky gold
scarf: pincurls for later.
You'd ask,
can you remember this shopping list?
and all the way to the store,
down heated sidewalks,
at intersections
waiting for the walk signal,
our hands
stayed clasped securely
as I recited to myself
bread, coffee, cigarettes, milk.

It was like this, then,
when the boys came
with their slingshots, sticks, rocks.
I thought the house, the palm tree,
were ours
but we were renters, and new.
The boys showed us
a custom we hadn't heard of: shoot down
the nests, watch eggs break
onto dry cut grass or
baby birds split apart,
fragile bellies spilling blue
and red intestines,
beaks still gaping
for air, rescue, filling.
I stood frozen
with hatred.
I think I screamed,
but did I?
Could I move?
I think I chased the boys, threw my own rocks
but did I only want to?
I remember the aftermath—
the boys gone gleefully
into the L.A. streets,
me falling to my knees,
handling the horrific remains,
burying small soft ones
in the lot next door, alone:
never telling you.

One day there was a survivor.
It lived through the bursting nest,
the long fall, impact.
Struggling in my palm,

trying to get to its feet,
beak an aching **O**
that called to every cell
in my female body: *feed me.*
I carried it into the house, to you.
You helped me gather the shoebox,
cotton, grass. We walked
to the store, bought a set
of tiny doll's bottles.
All the way home,
I thought I could hear cries
but when we arrived, silence.
We found grayish skin
limp against cotton:
dead from fear or shock.
We left the plastic bottles
on the counter.

Mama, I've had you back
all these years
and still I awake to the morning
clamor of birds
in trees, a particular cry,
not knowing how to satisfy that
open shape of need,
remembering how you tried.

What Part of Me

What part of me said yes?
What part of me gave consent?
What part of me motioned you forward,
nodded, spread my legs for you?

Was it my long black hair,
bangs uncut, tangles uncombed?
Was it my skinned knees, unbandaged?
Was it my thrift store clothes?

What part of me invited you in?

I must have tempted you,
been easy prey:
unwatched, unguarded. Seven years old.
That must be what
you took as permission
to put your body into mine
to put your fear into my gut
to put your anger into my mouth.

You stole parts of me:
legs, hips, what rests in between—
you took half of my body.
Chopped it up into parts I possess
and parts I have lost,
parts left alive
and parts paralyzed
parts in the dark

parts in the light
parts I can see
parts you made invisible.

You cut me up,
left me alive
but scarred, black holes
scattered
throughout the universe
of my body. You removed
the twin gifts of grief and pleasure
from my blood.
I want to know,
what happens now
when the edges
of flesh and memory
begin to awaken?

Wildflowers

Some flowers fold up
at night like prayers,
clasp petals around
a virtuous core.
Others gape open
through sunless hours
like eyes refusing
to admit they can't see
what scares them.
One flower dangles
her pink lips curled
out, or down—
already she is heavy
with tiny seeds.
They darken inside
her lengthening pods.
The evening wind
pushes at them
through the thin membrane
but it's too soon;
she won't split open.

Certain flowers have no faces
but thrust like hands
out of crumbly earth,
long white fingers
stabbing crazily.
They are compasses
who find magnetic North

in every direction.
And always there cluster
tiny round purple
gold or blue blossoms,
ones who rely
on massed roots
for existence.
They take pride
in their scent,
how bees come back to them
in the morning.

At twilight I walk
through that barren landscape
where nothing needing care
grows, where abandonment
suckles survival
from all that is tender.
I learn to step
off the dirt road,
cross the ditch
singing with crickets,
go into the tallest grasses.
Sharp seeds and thistles
cling to my hair,
bite my ankles,
welcome me in their
tangled, scrappy ways.

I don't know their names—
common or Latin,
real or imagined.
I don't know families,
phylum, species.
But I know how
they comfort me,
make a suitable path
for my heart.
When I forget my own name,
wildflowers don't mind.

Formula

with thanks to Grace Paley

I don't know where this story starts, or even if it is a story at all. There's no truth in the old formula of beginning, middle, end. Every time a person says, "Once upon a time," it's as if that person becomes an eagle, choosing to suddenly plunge beak-first into the middle of the ocean. How will she survive in a new element? Will she make it to shore? Which direction is shore? All directions? No direction? The horizon is misleading. No matter how beautifully the sun rises and sets, or how thickly clouds bed down low on the salty breath of the sea, the horizon keeps moving. The eagle is still there in the water. The problems are always the same for me with this story: backward or forward? My story, or my mother's? Win or lose?

I've been trying to swim out of the ocean of this story for so long. Finally it has occurred to me: I am not meant to find the shore. I am meant to dive below the sterile surface into a world where time doesn't travel in a straight line, but becomes one vast continuity, a thing as round and whole as the planet. The only problem with this plan is learning how to breathe underwater.

One way the story can begin is with a family. There is a father, a mother, a 2-year-old girl. They live in a nice neighborhood. The father, a handsome blonde man, delivers milk and dairy products. Imagine stucco houses, his white uniform, shiny black shoes, the truck waiting by the curb in the sun. The mother, her classic French/Jewish lineage glowing in a luminescent face, brushes her dark hair into waves and wears crisp cotton dresses to clean the house, stroll the girl, make dinner. Her waistline thickens: there will be a son born in September, but for now it's still late June, soon after the Solstice—hot and bright.

When the girl is found unconscious in the white tiled bathroom beside the heavy porcelain sink, claw-footed tub and precisely hung towels, this story falls rapidly into conclusion. The aspirin bottle empty on the floor, the little girl's dress, white and pink, belled out around her still-chunky limbs, the way her eyes are closed, yet sleep is so far away.

The end. Let this story sink to the bottom, settle into a deep crevice, a thick layer of decayed plant life, a level of blessed darkness. Make a new story; that's how the formula is supposed to work.

But the story only begins again.

The mother remarries. She remembers the baby she lost and tries to replace her; in fact, she has another girl. Everyone says this new baby looks just like her lost sister. Once again we have a father rising early for work. This time he slicks back his black hair with Tres Flores and pulls on steel-toed boots, heads out at dawn to the construction site. The mother, still beautiful, still in cotton dresses, cares for her new baby. She learns to make tortillas, chili verde, sopa, speaks Spanish to her new neighbors. They are a family. Aren't they? But why hasn't the father come home, as he is supposed to everyday at 5:00? Why is the pot of rice left burning on the gas stove? Who has left this child unattended, the two-year-old girl crying in her white dress and snug leather boots, sitting in the corner of her crib? Haven't I already told this story? Is the ending so different? Please, let this story be swallowed by ancient waters; let it end.

I can't stop beginning this story. Somehow, that girl survived abandonment, grew up, married. Once more, there is a father, a mother. There is a girl, age 2; her brother has just been born. The little girl is asleep; it's a hot June night in a small apartment. The father is in one bedroom; in a few hours he'll have to get up to deliver papers before returning home to shower, kiss his family goodbye, leave to teach all day. In his dreams he smells newsprint and brown rubber bands, the soft river scent of the valley before dawn. He hears the tick of his alarm clock, the passing of minutes before he needs to get up, start making money again.

In the livingroom the mother is awake, walking the new baby, who won't sleep. She thinks he is hungry. She sits down in the rocking chair, winces as he latches eagerly onto her full, sore breast. For a moment there is silence while the baby sucks hard, swallows. But then his thin, heart-broken cry comes again. The mother's body breaks out in a sweat. What's wrong with this baby? What's wrong with her? Why does it hurt so much to nurse this ravenous child? Why can't he get any milk from these breasts? What should she know how to do that she doesn't?

Despite exhaustion, the woman knows she won't wake her husband, won't ask for help. He is giving her all that he can, and it is not enough. This child is her responsibility, her work, her own darkness to hold and feed.

The baby has a rash from the heat, and from incessant crying. His blackened umbilical cord stump pulses in and out with each scream. The neighbors on both sides and below toss angrily on their hot wrinkled sheets. In the second bedroom, the little girl stirs in her bed; often she doesn't sleep through the night. On the livingroom wall are a hundred shadowy hands moving and overlapping in the sway of the young maple tree outside the window. *I can't do this,* the mother thinks—inside her heart like a scream, but silent, and uglier.

She paces the small path from front door to kitchen, back again. Her hair is long, needs to be washed. She has it tied back. She wears cotton pants and a baggy T-shirt over a heavy body that was C-sectioned only a week ago. She has always resembled her father, until tonight. Tonight she looks like her mother, who walked and rocked sick babies, whose desperation found solace in purple wine, who lost her babies to accidental poisoning and foster homes.

This is a hard story to swim through. To tell it, I need air. Let me be a sea turtle, then: let my eagle's wings transform into huge strong arms shaped like crescent moons to pull me up, and give me heavy back legs with webbed toes to push against water, and let this sharp hooked beak stretch up to break the surface. Let me tell it in another form.

The baby falls into a light, restless sleep. His brown rashy skin sticks to the mother's arm. Moisture collects in the spaces between their bodies and the mother's breasts harden and burn like two suns that will burst any second. She rocks her baby, rocks and rocks in rhythms emanating from her bones. Into that tenuous calm enters her daughter's steady breathing as the girl sleeps on, all night, the first time in two long years. The mother sees the maple leaves outside make the crucial transformation from gray to indigo to rich dark green. The fine edges of shadows are tinged with white gold sunrise, and the air changes too—rises from the cooled earth like a clean, quiet word. The mother thinks of the cans of powdered formula sent home from the hospital, how hungry this boy is, how she sees only now what she has lost. This is the night the woman begins to understand and open to a kind of desire: she wants to name loss, own it, display it fiercely.

This is the story I am afraid to tell. This is the story I crawl out of in my borrowed turtle form. I don't know where it starts. I only know I can't let it end here anymore.

Venom

Through heavy thick sleep
I hear voices: my daughter,
just seven. *I had a bad dream . . .*
and my husband's practiced reply:
Crawl in here with us, babe.
Miranda snuffles in under our blankets
wanders down
a windy path
*bees . . . all these bees
around me . . . stinging me . . .*
she falls asleep
sharing her father's pillow,
safe.

    I dream too:

    a man's fingers thrust inside me
    there's not enough room
    it hurts
    I am in a sleeping bag
    on the floor
    next to my girlfriend
    he touches her next
    we lie silent
    clutching each other
    afraid to cry out
    we both pretend to sleep
    we can't be roused
    we can't help each other

it's dark except
for the light on in the hall
all I see is his curly head
moving beside Hannah
and a blur of white skin
as his body
moves in the narrow space
between us.

In the morning—
there is a morning—
my underpants are gone
I wear just shorts and T-shirt
to pick apples
up high
he stands beneath me
to balance the ladder
his hand
slides under the thin fabric,
up my leg, no distance at all
for his fingers
I want to go home
but he is the adult
he can drive
I'm only seven

and Hannah's freckles are dark rust
on her pale cheeks today.

The apples are small
and falling into deep wet grass.
Yellowjackets buzz thickly,
eating into soft
bruised skin.

Careful, he says when I jerk away,
You might get stung.

In the morning my daughter wants
cartoons and hot chocolate.
She won't remember her bad dreams.
She never does.

Certain Scars

At night
a wound throbs
bitter and lonely
under a thin membrane,
cries against silence.
You must sit up with it,
rock phantom pain
in your arms near
a closed window.
The black sky looks in
and makes no comment.

No one asks
exactly where or when
it happened.
You don't say, *This
is where a living part of me
was hacked off.*
You manipulate your body
and other people
in crowds and tight spaces:
you avoid touch.

If beauty is symmetry—
lips, eyes, all limbs
in balance—then
you are ugly. The wound
migrates through muscle
and bone. It shows up raw

on your face in a photograph,
grows on your breast
under a lover's mouth, gapes
open in your hand
held out to a friend.
This scar will not share you.

Lost Language

with thanks to Marie-Elise Wheatwind

Can you read the word
that is written on my forehead? Maybe
it is stitched in tiny crimson beads,
flowing like Hester Prynne's 'A'—
gaudy, elegant, brave. Maybe
it is slashed in spray paint:
a name—a code—a lie.
Maybe it has always been there
and I never noticed. How secretly
wrinkles and lines must have grown
into that language long since forgotten,
a word my great-grandmother
murmured in her sleep, traced
on her own forehead with trembling fingers.

The sound of your voice, full and strong,
reading this word out loud
becomes the sound I have been waiting for,
the sound worth everything, so sweet
I hate you for what it reveals.

Some believe the lines of a hand
predict children, life-span, wickedness.
Does this word on my forehead
tell you there is an animal
pacing in my heart—
how she watches, cries, beats
my rhythms to her own?

Does this word warn you
she is twitching her long tail,
waiting to be freed?

Read it silently at first, like Braille
beneath your fingertips. Touch
my eyelids so I can see its shape,
stroke my lips so I can taste
the bitterness, press your palms
to mine as if they could speak.
I promise, I will say it,
that word,
no matter what language,
no matter how obscene—
won't I remember all
they say I have forgotten
when you read the word
that is written on my forehead?

Finishing What He Started

After he leaves or I escape
I hold myself together
with metal staples
or stitches of sinew,
or my own bloody fingers
because I know
I am not strong enough
to let go.
I hold on, hold
together edges of a gash
he made with his body; I can't
split open, can't fall
apart or give way
not now—maybe later
not now—maybe never
not now.

But this time metal rips
sinew snaps
my slippery fingers lose
their grip,
I crack from the center out
belly tearing like
sudden lightning, a ragged line
ripping up from my navel
to my breasts, splintering
ribs, slicing my throat.

Down through pubic hair
and bone to vagina
I am splayed open
like a starfish, a dead crab
an ancient turtle whose shell
has finally been broached,

and what comes out?
What unseen thing
did he leave inside me
festering, growing?
What emerges
from this unholy rent?
I won't look
I won't.
I
look,
leaping clear
covered with my slick warm
fluids, glistening red and wet
a whole woman stands
breathing, panting.
She didn't ask
to be created
she glances back at my open body
does she think *wound*,
or *birthplace?*
Does she wipe off her face
and walk away
or turn back, kneel,
hold me?

Listen,
something strange:
I think it is a laugh
or a cry,
or a lullaby sung
in an unscarred voice.

Bodies of Water

Sometimes the Open Hand of Desire

slaps you right in the face, hard
enough to leave a palm-print
etched on your cheek. Sometimes she cups
your chin, runs a finger along lips and bone.
If you are lucky, the open
hand of desire trembles
warm with the power to please. If
you are unlucky, she cuts you
with her sharp fingernails. But at least
once in your life, the open hand
of desire should stroke
your hair back,
wipe tears from your eyes, hold
your face firmly in her grasp
and guide you into her kiss.

Refuge

Everywhere we meet that spring, rain—
in pollen-dusted pine woods; in your car

parked beside a swollen dark river, under trees
dripping limp cherry blossoms. Even

in our small haven on the hill, gray sheets
come down in pearly sound, steady rhythm

against words that cauterize
our raw, wounded mouths—rain

to cleanse us, seal us, witness the fluid
rebirth of laughter; rain

to calm the furious pace of discovery, write
the vows we swear, our bodies bare like converts

receiving true religion. We touch
as if nakedness can save our souls, as if

a caress is the last honest lie left
in this war. Skin on skin

rain on leaves. We find the cool path
through fire's destruction, let rain

drum ash and bone into black earth,
awaken the seed of a miracle.

Summer Solstice

Your hand still holds mine, clings to peace
rarely reached in daytime. Humid air
drifts in the window, tiny breath of summer
on the heels of last night's storm.
Like extra limbs, sheets twist around us,
with cool, smooth scent. I breathe
slowly to make it last, trace
the silver circle on your finger, feel how
it has grown deep into flesh. There's no telling
where precious metal ends, where you begin.

We celebrate this victory each year, relish
our right to lie here, listen to muted movements
down the hall as children wash ashore
from sleep and old foundations marry into earth,
relax into a new century. I wonder
if this ring will one day fuse
to bone, synchronize with pulse
and nerves, slowly merge elements—
so native it becomes invisible, absorbed
by the power that is you, man who sings
in my blood, whose hands heal me like a blessing.

Bodies of Water

The rains continue into dawn, into daylight
and the air fills with water, heavy
molecules of hydrogen and oxygen
clasping each other tightly.

Our bodies swell with dampness,
pores breathing it in, wet, warm.
This rain enters us, soothes spaces
left dry by things we've lost.

"We're water up to here," my son tells me,
flat palm measuring floor to chest;
his interpretation of a science fact
learned at school; but what if
he's right? What if we are vessels
mostly filled with clear, sweet water?

Then when we walk, we splash and ripple
in a liquid rhythm. At rest we are pools
of crystalline silence. When I touch you,
sweetheart, rivers surge between us, flood
full across our fragile, tumbling boundaries.

For Shame

Shame is a little girl
who hides
her face
in the shadow
of her absent mother.
Even the act
of lifting
her hands to cover her eyes
is too embarrassing—
what have those hands done,
or not done,
what are they accused of doing?

Shame doesn't cry
but wants to, doesn't tell
but is supposed to,
won't reveal herself but sees
she has already exposed
guilt with her awkward gait,
her slumped shoulders,
the sullen softness of her voice.

She doesn't believe anyone
still loves her, knows
it's true what the boys say about
her lost in the woods;
worse, how everyone else knows
losing her way was her fault.
That they were men,

not boys,
and she's only 7,
doesn't matter.

Shame isn't pretty,
she's growing too fast
and has hips
and breasts, smells like sweat,
like coarse hair in crevices.
Her clothes don't fit right,
need mending or washing, or maybe
that's just because she hasn't
really been clean since
her birth.

Shame is the kind of child
women have and want to disown,
secretly praying: God, grow up,
move away, change your name!
Her siblings want to
forget she ever shared
the same small house.
No proudly framed pictures
of Shame's brief childhood
will endure.

I want to write this poem for Shame,
dedicate tenderness
to her burdened, unloved body.
I want to touch her dark face
and turn it toward the sun,
make her stand tall.
I want to kiss
her clenched palms and fingers,
tell her she has
beautiful hands.

Commencement Bay

A child's swing lifts you
out of your body on the swell
of a wave that starts in your womb
rises up through spinal cord
and blood like a long shuddering sob

your hair coming loose from its bindings,
streaming out in a rippling black comet
with cold fingers of light
and salt air sliding
along your scalp, tangling
and untangling, knotting and unknotting

your hands melding to gray links,
knuckles and bones fused
like cramped chains; the letting go
is your only fear,
your last hope.

The Territory of Love

The territory of love is a new continent always just being discovered by those who believe in it, even as myth. You've been taught this land was lost, like Atlantis, or the secret passage to cities of gold.

Yet the territory of love is what you are driven to conquer, what makes paved roads too tame and distills wild courage out of pain; this territory is what you realize as your destiny on non-existent trails with panthers crouching in wait at every watering hole.

There is no end to the territory of love. The deeper into the interior you go, the more surprises await. There may be natives sauntering along naked, passionate hearts exposed and unashamed; and tender songs piercing the civilized weave of your cotton and wool clothing. This is their land. You are the stranger. Customs spring up like unknown plants whose fruit you must taste but not swallow till the juice proves palatable or poisonous.

Along the coasts of this territory even the seabirds have strange voices full of longing and the oceans lap up on so many kinds of beaches—black rock, smooth white sand, roots of tangled forest—you think this territory transforms itself even as you stand there rubbing your eyes, dazzled and terrified by your own fierce desire to enter this land and make it your home.

Beware. The territory of love does not take kindly to being colonized. To live there, you must give up what you think you have learned, take back what you fear is forsaken: understand, you can only win by losing.

A Walk in the Forest

Enter me. Become lost in me. As time passes,
learn to be wild. Venture off trails, live
on bright berries no one else has tasted.
Slip through thickets of swordfern,
salal, cedar; feel the green pulse of my heart.
My breath hums along your cheek. Listen.
Day and night, I'll surround you with songs
of passion sung in proper season.

They told you this forest was haunted by ghosts
of a crime never laid to rest. It's not true.
Yes, old roots anchor deep in survival,
yet my soul is newly leafed: tender, turning,
opening toward light. You'll have to take
your chances here. Stay long enough, drink
from these cold, spring-fed waters:
you'll transform into a creature
whose delicate tracks don't own,
but possess.

Grief

My ribs ache,
bruised from the inside
by a pounding
of questions,
each breath a fist,
a desperate demand
for release.

I clean, do a wash, feed the cats.
I drink coffee.
I take down
children's artwork
in the kitchen—
nine months of paintings,
awards, stories.

I lean against the wall,
let sobs come deep and harsh.
I press my cheek to
crayon drawings
whose slick blues
and greens resist
my sudden grief.

I look at the spiraling
construction of my hands.
I read a book,
wash my hair,
let it tangle in my fingers.

I wonder,
what have I unearthed?

I remember an embrace that has gone on
into the bones of this world.

Sorrow As a Woman

Sorrow is a strange visitor. Where does she come from, anyway? Out of nowhere, out of your future, out of ancient lands you didn't know she walked. She comes into the room where you are sitting at your work, and says nothing. Her legs are bare, her feet are dusty. She wears an old cotton dress the color of dark poppies. Black haunted eyes stare out at you from beneath hair ragged as burnt pine. Sorrow stands too close beside you, and the scent of her loss is sharp in the air, your mouth. It brings tears to your eyes, thickens your throat so speech is impossible.

Sorrow compels you to fall to your knees in front of her and pray for grace. How will you house her? What bed can she sleep in? What exotic foods will she demand? How will you entertain this difficult guest, and when, oh when will she leave? You don't even know what to call her, so you make up names: *My Sorrow*, you say. *Dearest Sadness. Sweet Grief.* And in time, Sorrow opens to you, tells you stories of her travels, what led her to your doorstep. They aren't pretty tales; you expect that, given the roughness of her hands, the way her bones show so clearly beneath cheeks, shoulders, hips. But you are surprised to see a smile hover around her dry lips one day. *Ah, there was a place,* she begins, *Beautiful, magic woods* . . . She stops, touches your wrist.

You remember the young forest rising up, thick and green, the small creek flowing by in its clear cold bed; the way sunlight illuminated veins of wide maple leaves. A strong force sheltered the spirits of lovers under these trees. A holy ground was created . . . but the lovers have moved on, their spirits have moved on, and the creek, the trees, the verdant lands are destroyed. You jerk away from

Sorrow's clasp, angry that she knows this. Then you see her face: smooth, quiet, eyes closed. Deep in her tired body lie the fragments of joy. Sorrow has nourished them in her rich memory.

Now you know what to do, how to pray to Sorrow. Bathe her in water whose source is pure. Stroke scented oils into her parched skin, comb out her hair with your fingers. Take her to your bed and taste the bittersweet loss of her mouth. She will be your most intimate lover. She will never leave you. From her womb she will birth heartbreaking beauty. Whisper to her: *My Lost One.* Then she will call you by your secret name.

A Ceremony for Crying

1. Invocation

Cry for hunger that feeds on betrayal,
a serpent devouring itself.

Cry for fear whose belly brings forth
misshapen anger.

Cry for grief swollen into
the religion of denial.

Cry for lost chances, missing
like holy birds from the sky.

Cry for long knives of separation
and abandonment, unsheathed.

Cry for burdens clinging like
parasites, without pity.

Cry for love like wind,
existing without body.

Cry for empty hands moving to
empty mouths, again.

Cry for escape without homeland,
exiles with no names.

Cry for hatred worshipped,
made into human flesh.

Cry for ceremonies with *no*
the only song left.

2. Naming

Cries like music that cannot be heard
in a roomful of ears.

Cries like poison hidden
in the breastmilk of strong women.

Cries like criminals, their open hands
deviant and unacceptable.

Cries like bats rising from caves
in a deeper age.

Cries like trapped pheasants
bursting cover.

Cries like blood that will
not be washed clean from stones.

Cries like late snow
that melts with no trace.

Cries like sounds never voiced,
a tongue cut out.

Cries like pain stretching
the length of a body.

Cries like mission doors
slamming shut.

Cries like the fierce heart
of the earth breaking open.

3. Blessing

Cry for birthing, how power enters
and leaves at once.

Cry for astonishment, given to us
like stars above a dark road.

Cry for silence, the silhouette
of deer in morning meadows.

Cry for trust, a yellow flower
opening in black lava beds.

Cry for remembrance, like bones
undisturbed in the sweetest earth.

Cry for understanding, a sound
like water pure from the source.

Cry for reunion, the way miracles come
even to survivors.

Cry for desire, the path to the center
of an ancient mountain.

Cry for wholeness, a dream within
a dream you are given to praise.

Cry for strength, how it spirals
like smoke carrying prayers.

Cry for faith, a sacred plant
ripening with thick rich leaves.

Heartwood

A hard, fast wind rises overnight. It comes in sudden rushes, pushes against the corners of the house, wraps itself around the eaves and roof—surges—then, it's gone. Where does it go? Where is it coming from? The house braces for the strength of wind's arrival—finds itself alone, untouched, tense against nothing but the warm wet breath of Spring dawn.

In my bed, in the dark, I think the cedars along the road will give warning in the tops of their feathery thick branches; I try to pick out the rattle of the big dogwood's long twiggy fingers, or the heavy song of my old cherry tree's strong voice. But there is no gentle beginning to this wind, no slow swell that builds predictably into well-known roar. Only the restless silence of earth curving over a bay on Puget Sound very early in the Spring. I listen. Hear how this place is subject to moon, tide, currents that run cold down past islands in the Straits of Juan de Fuca, or seep warm with scent from humid tropics, spreading dry and heated from the Gulf of California. In this way trees recognize wind coming while wind is still in its infancy, taste wind in their roots, a difference in the way dirt parts around their blind reach, movement in some continental shelf, a wave of moisture from mineral springs or a salty tang from deep in the passageways that lead to ocean.

Yes, lying here in the blue-black light from my bedroom window, I am sure that the trees know this wind, everything about it that I don't; but they aren't telling. Or are they? I touch the old lathe and plaster wall above my head with my fingertips, feel the long pause of house timbers. Old-growth beams far down in the basement are talking, creaking. Faint messages swirl around the roots of my house. Heartwood remembers, gathers itself for the next leap. I close my eyes just as the whole world takes wing.

Vernal Equinox

for Margo

Your body is so filled
with visions, your fingers ache. Outside,
tender tips of leaves
flare on the lilac, rain falls
slow and warm.
The earth swallows.

Veins in your hands
quicken and shine
through dull skin.
You are thin, hungry,
your body's sweetness consumed.
Though it hurts, you write to me:
Rivers are breaking up, huge chunks of ice
float past.

All winter in our separate
battles, we've waited
for a sign,
a map, revelation shining
in the air. We want a miracle
like our ancestors saw,
women who must have once walked
together in the same wilderness,
knew how to pray
in the frightening language
of abandonment.

No messages come to us
except the ones we write
for each other
saying, *don't give up,*
be brave,
hang on.

But this morning, stars
change places, the equinox
arrives in darkness. Listen
for a long time
to the piercing song
of a bird who loves
this dark hour.
Her wings hum a note deep
as unanswered desire.
The path of her shadow flickers
in trees like a glimpse
of suspended writing.
I send the words to you,
imagine you tracing
their silver glow, cursing, but
letting that heat brand
your fingertips:
we're alive, alive
alive.

Going On

In the morning my daughter knocks
on the thin door of my study.
She wants to be let in.
I put aside work and greet her.
She sits on my lap, almost too big
and only eight years old.
Her hair is rough, unbrushed, but clean.
Underneath strawberry shampoo
floats a musky, pre-pubescent odor.
I feel the weight of her thighs on mine
and her sturdy arms firm around my ribs.
This moment may be the only sweetness
found between us today.

Earlier, at dawn, I'd stepped out
into the backyard, looked up.
Stars pierced the crisp blue illumination
of sky. Cherry tree leaves hung dry,
still, poised. I remembered what a friend
told me: that trees inhale all day,
exhale all night. What a long breath!
And in that cusp between *gather*
and *release,* I imagined anything was possible—
a deep rhythm going on into its own history
like a masterpiece of song
or the concentric ripples of echo.

Now, holding this girl's body
full to bursting with future,
I want to know everything—if music feels pain,
if echoes experience awakening—if,
in transition, stars celebrate
the inevitable journey
from summer to winter constellations.

The Night of No Shadows

There is no dark to soak up light;
the gentle beams of streetlamps bounce
against glowing cloud cover,
stream back to illuminate
smooth white cars,
sidewalks, trees unable
to sink into shadow.
Our eyes are not equipped for such
darklessness.
This is not lunar or celestial
spark;
this is the eerie radiance
across the threshold
of dreams. My son wanders ahead in tall
black boots, churns a new path, his head turning,
searching for the source. Small
against the muffling snow, his voice floats
back to me, full of a three-year-old's persistence:
Is it angels? Is it stars?
Is it the moon come down from the sky—
broke, spilling all over the world?
I try to answer but my lips move slowly,
bathed numb in cold luminous air. My son stops,
flings his arms open, palms up, laughs. I laugh too,
though it hurts my teeth,
my lungs, my heart—
I see the source of vision, of stories,
I see my son's body
formed to receive light.

Winter Solstice
for Daniel

Years of darkness, I listened to rain. Years of sitting in a chair, lying in a bed, standing near a window listening to rain, watching pine trees grow wet and heavy, maples mute and soft with dormancy. I didn't know this moment was ahead of me. I didn't know how I would be changed, how everything I knew about myself would be bared in lightning, how burnt I would become, how cleansed, amazed; least of all did I guess I'd come back to this place, this rain, and sit here transformed, grateful. I didn't know about loving you, loving myself, betraying all that. Still, the rain can't be different. It can't have changed. Yet I hear it more cleanly, with less interference. It is not a language once known and forgotten. It is what comes before language, or after language is exhausted: a deeply aural text, a rhythmic, flowing, drumming, trickling manifestation of alchemy. Magical coalescence: moisture, air, warmth. Falling.

Some things don't need to be written in ink, or blood, or even on my heart. Rain inscribes a black asphalt highway, wide green leaves, a path for my feet, with symbols that can't be translated—only danced, rocked, hummed. I've come back to rain as if coming back to my infancy or that immeasurable existence before birth when I grew toward those years of darkness, and you, and all the years coming now wrapped in wordless, clear, resonant downpour.

Three Poems for April

1. Lilacs

The way the bush has become a tree
fills the window over this sink,
makes doing dishes feel like surging inward
into a younger place.
Maybe it's the rising
face of the east reflecting green cells
in heart-shaped leaves, the tight purple
glow of lilacs: cold, but fresh.
I know the clouds and choppy wind
from a secret place in childhood; they cover
what winter-shy eyes can't bear.

In such a moment my hands linger
warm and slow, want the slick patterns
of cup, plate, knife. The bend of branch
with five honest leaves prepares a sentence
for my daughter's ears: *A little girl
was raped.* (She will frown, ask,
What's raped?)—or maybe like
the half-open petals, close and rushed:
*from your school/a girl, walking/someone
hurt her.* But no words account for gold
caught within a bush, the clarity
of wet violet—oh, this light.

2. Rain All Night

Rain all night and I dream of rhododendrons, white and crimson; tulips not yet open in streaky orange. A woman asks me to gather bouquets for vases on tables, arrange branches of starry dogwood. I wander from woods to garden and back again, arms full of a crazy mixture. Petals beaded with water wash my bare arms, darken the cotton of my blue skirt. I keep saying *oh this one, this one.* I am expecting an end to beauty; but there is always more, more. My own surprise blooms bright against despair.

3. Survivors

Like animals we bare our bellies.
You slide the length of my body,
press a rough cheek against my breast,
asking wordlessly
for the same comfort
you've given me all this dark month
when my body withdrew into nightmare.
Now I want to hold you like this:
in daylight, arm around your shoulders,
my heart singing slowly in your ear.
You fall asleep, muscles moving electric
deep within your arm.

I don't sleep but watch the glitter
of new leaves outside
rippling like a threshold
between memory and denial,
begin to trust how void, light, breath
all balance, make the center
of wound become world.

                    You awaken,
say, *We've been through something,
haven't we?* as though you are just
now shrugging off the skin of a dream.
I lean into your kiss, tell you
you are mistaken,
that we've only taken
the first step
into a tenuous truce
with history. Still,
you stay,
accompany me toward the future
I can barely see.
Buried in the past I came from—
violation, fear, a distrust
of the body—waits
a season I will claim
as my own sweet time.

Indian Cartography

Naming the Nameless

In the Straits of Juan de Fuca
the islands spring forth
too many to count.
Covered in dark green pines,
rising straight out of volcanic dreams,
round and steep.
They hover on blue waters
like bristling turtles or rolling
leviathans caught
surfacing—
held, clutched, in some sudden spell.

One island grows differently.
Gold, smooth sides swirl
with tall dry grass.
Evergreens punctuate a sinuous spine.
We pass its length slowly,
watch waves crack
against rocky edges.
Black-tailed deer graze
in a hungry grace. But I swear
a few of these animals
gleam solid white.

Along the high
crown of this island
comes an eagle. Such a wingspan
needs only easterly winds
and inborn balance

to keep pace with our progress.
There are levels of beauty
that tear my heart
from my chest: Sky,
eagle, pine, deer,
grass, rock, Sound.
Each one a canopy
sheltering the next.
Each one a strong
back that holds up
the one before.
Going on—going deeper
than I can bear to name.

I Am Not a Witness

I found Coyote, Eagle, and Momoy
in a book, but cannot read
the Chumash words. I found
photographs of bedrock slabs pocked by
hundreds of acorn-grinding holes,
but the holes are empty, the stone
pestles that would curve to my grip
lie dead behind museum glass.
Mountains and rivers and oaks rise
in Spanish accents: San Gabriel,
Santa Ynez, Robles.
These are not real names.

Some of our bones rest in 4000 graves
out back behind the Mission.
Some of our bones are mixed into mud
to strengthen cool thick walls
where smallpox and measles came and stayed.
Some of our bones washed down the river
whose name I do not know
past islands I cannot name
to the sea where
I have never sailed.

Mixed-blood, I lay claim by the arch
of my eyebrows, short nose, dark hands.
I am not a witness. I am left behind, child
of children who were locked in the Mission
and raped. I did not see this:
I was not there—but I am here.
Where is the place that knows me?

Without History

for the Woman of San Nicolas Island

Once I dreamt that the truth was inscribed
in bone, sacred skeletons waiting to be found—
messages translated and sung out
in a genealogy of memory.

Once I believed my account survived, written
on my heart—a secret fragment
carried safely to some future place
where blood is ink, like faith,
indelible.

Once I trusted our story to my tongue:
told it to my child in milk-language,
first sounds of a dialect
woven from the certain web of the past.

Now you see me as I am. Alone.
No trail to follow back to
a genesis of soul.
Unable to tell what I lost.

They call me *survivor*, but
there is no honor in what I came out of,
no joy in a testimony of ashes.
All those who knew me
fell into extinction.
My history
abandoned me in smoke.

I've sifted the earth for bits of stone,
a lock of black hair.
Nothing remains—
only my cupped hands
like burnt baskets
too empty to hold a cry.

Indian Cartography

My father opens a map of California—
traces mountain ranges, rivers, county borders
like family bloodlines. Tuolomne,
Salinas, Los Angeles, Paso Robles,
Ventura, Santa Barbara, Saticoy,
Tehachapi. Places he was happy,
or where tragedy greeted him
like an old unpleasant relative.

A small blue spot marks
Lake Cachuma, created when they
dammed the Santa Ynez, flooded
a valley, divided
my father's boyhood: days
he learned to swim the hard way,
and days he walked across the silver scales,
swollen bellies of salmon coming back
to a river that wasn't there.
The government paid those Indians to move away,
he says; *I don't know where they went.*

In my father's dreams
after the solace of a six-pack,
he follows a longing, a deepness.
When he comes to the valley
drowned by a displaced river
he swims out, floats on his face
with eyes open, looks down into lands not drawn
on any map. Maybe he sees shadows

a people who are fluid,
fluent in dark water, bodies
long and glinting with sharp-edged jewelry,
and mouths still opening, closing
on the stories of our home.

Migration

Five a.m., at the burnt end of a fierce October.
It's dark, quiet. I wish I were on the road,
traveling, a thin paper cup of coffee
hot in my hand, dawn coming up
on the horizon and all around me
the outline of shaggy pines, uncut pasture
flashing past car windows.

I'd go north into the foothills,
to one of the passes—
Snoqualmie is open, I think—
over the Cascades and down
into the flatlands of eastern
Washington. I could be there by noon time,
walk the arid soil beside the Columbia.
I'd sit in a lonesome sandy spot, listen
to the restless water, smell the scent
of departure.

This time of year, birds gather—
Canada geese in clouds of gray and black,
herons stick-legged under generous blue cloaks—
and my body, ready, waits for the moment
when wind insists, sunlight strikes
a true angle, triggering in my heart
some secret knowledge of direction.

January Cusp

Behind houses with no lights
and trees without leaves
the moon rests on her curved back.

She spills onto cats
crouched on porches,
the smooth black fingers
of scavenging raccoons.

Are you awake?
Do you see the moon rocking
like a cradle
under the touch
of a dark mother's hand?

Correspondence
for Janice Gould

I spend the summer near water:
McKenzie River, Guemes Island, Birch Bay.
You spend the summer driving
from Albuquerque to Santa Fe, back again.

I try to teach my children how to swim.
You try to raise poetry in adolescents.
In the evening, crickets scatter
in driftwood, sing me to sleep.

You write of mourning dove skies,
cicada's hypnotic buzz.
Despite distance we are sisters
birthed from the same wound.

Half-breeds, we don't know
our own names, or stories.
But we can recite the creation myth
of race, species, inferiority.

We can remember there are others
like us, unfound.
They have a certain darkness of face,
eyes, thought; they move silently

through dry desert towns
and down the long Pacific coast.
Maybe they are looking for us.
I'll leave a small pile of feathers,

shells, and round black stones
as a sign that I came this way.
In your classroom you do the same
with students' words, hearts,
                      jagged-cut poems.

After Colonization

the land divided her loyalties between native
and foreigner, then and now. We fell
between the cracks like a laugh cut short,
or a sob. We can't forget what it was like
to be her first, her only, even as history encodes
our bones with change. Whether by rape
or love, violence or choice, we are survival
made flesh. We walk through life unshielded

and find boundaries, treaties, reservations
that don't speak our names. We make camp,
make dinner, make love. Make war. We look
for our own land to claim. When we find it,
we'll declare a holiday for our children
to celebrate. They'll learn ritual songs,
memorize an arbitrary date, carve petroglyphs
of the first Half-breed to set foot
on the unfenced territory of the heart.

ghazal 8/7/94

Salt water licks at my feet.
Fresh water falls on my bare cheeks, arms, legs.

By nature of my species' evolution
I have a high percentage of salinity in my veins.

Yet I cannot drink this ocean—
inland waters nourish my cells.

My fingers follow the grain of a log.
I will sit on this shore the rest of my life.

Part one thing, part another.
I crave the depths where there is no division.

I will shrink into an old woman
who looks Indian and chants Hebrew.

When I have grandchildren, they'll celebrate
Simchat Torah, wave flags with blue stars.

Maybe I will take them to the sands
at Nisqually, or the mouth of the Klamath River.

Maybe it will be San Francisco Bay, Santa Ynez,
anywhere there are estuaries, mixing.

They feel the blood calling.
They will they know what it is.

When I'm gone, they will remember me.

For My Other Grandmother

for Marquesa Robles Miranda

You are the one I hear about when my father
drinks too much, the one I look like
when my hair is long and olive half-moons
curve under my eyes.

Marquesa, the one with four sons: the first
was for your husband—to take his name, follow
his work, be in his image—big, quiet, simple.

The second son was for no one—quick and laughing,
smooth as butter, you couldn't hang your heart
on him—in his eleventh year, a truck killed him.
You finally understood he wasn't yours to keep.

The third son was willful from the start—
you waited out a howling November storm
on the reservation, only an old Indian midwife
and the Sonora hills to catch that baby—
dark-skinned, strong, fighting already,
finding his way into the heart you thought sealed shut.

And the fourth son—he was for revenge—too tall
and handsome to be a Miranda, he was recompense
for the son your husband let slip
to that Valencia woman; you wanted to show him
you could grow boys without his help.

Four sons, Marquesa: one was my father.
Because you died so young, the only way I know you
is by the wide square palms of my hands, the long
"Indian teeth" of my bite, the widow's peak
of black hair with the stubborn silver streak
that stands straight up like feathers:

I know you by the way wine sings to my bones
when I am sad, when I grab my son, kiss his fat cheeks,
feel his hard head butting to get loose—
his dark, round body running away with my breath.

Gone Dancing
for Thomas Anthony Miranda

My grandfather never spoke Spanish.
I don't talk like a Mexican, he'd say,
nor dress like one neither!
Every day he wore clean chinos
and a fresh-pressed shirt to work
at the construction site, and for a night
on the town, his silky suit and tie.
In black and white photos, he poses—
solemn, dark, proud of his good taste.

But sometimes my grandfather disappeared,
was absent for days. His crisp shirts hung
neatly in a closet, his suit limp in its bag
from the cleaners. *Gone to the hills,*
people said quietly. *Gone dancing.*
Nobody remembers where, now. Some rancheria
that's vanished, some place so remote
only Indians wanted it. But my grandfather,
he knew. He knew when to go, how to get there.
Drove away at night, not a word to his wife,
his sons. Didn't take his good clothes.
Didn't tell where he'd been when he finally
came home. Still, word got around.
They said he was the best dancer, even
made his own regalia, wore feathers, shells,
grass. Girls fell for him, they said,
admired his legs corded with muscle, capable
of keeping step for days. Maybe that's why

he never took his boys with him. Maybe
that's why in the family album there's only
the kind of clothes all the dandies wore.

When my grandfather died, fifty years later,
his sons gave away closets full
of carefully creased slacks, clean shirts,
outdated suits. They didn't find
feathers, or shells, not a wisp of tule.
There's only this story, and my own
smooth legs, bare with desire to step and slide
certain times of the year, certain nights.

Riding the Back of the Universe

This
is how a
  path changes:
    The high rock walls,
      hardened clay and stone
      ground, fixed constellations
    which your eyes know in ancestral
longing all are lifted—shaken, shrugged
aside. You find yourself on the back of a
selfish, smelly animal who has had enough of
your burrowing into her spine, her neck, worrying
at dead ends, no sense of solution. With an inarticulate
curse from her throat, the Universe pauses to scratch. And
this path that is worn smooth from your callused feet,
from your hands and knees crawling, the long trail
of desperate prayers bumping beside you lifts,
shifts in the direction of creation. And
there you are. Saved. You didn't go
through, under, over, or around.
    You didn't move at all. This
      is the trick at the heart
      of impossibility:
  hang on past all
    endurance, and
then, let
go.

What Is Possible

Like clouds this evening when lightning
spreads sharply across the east,
stretches high into the west—
I am full of brokenness,
scattered in dark billowing masses.
A seal has been broached
and what was whole is cracked
and what was bound can no longer
find the healing thread.

I evaporate in this heat,
traveling without sound into the sky
or melting back into black water.
I am the fog on the cheeks
of women and men rising early
to find work; and I am the dew
glittering on the uncut blades
of grass. It is not possible
that I am forgotten.

What is possible comes at midnight
when the same jagged clouds
part, and a vast ocean is revealed:
stars, glistening like phosphorescence
left in the wake of the First People,
and a fearless moon, waxing gibbous and serene.
Blue, silver, luminous with torn edges,
the Universe goes on without our knowledge
or permission: a violent
and tender perfection.

Tehachapi Night
for Lisa Ohman (Elise)

Stormy. A hard wind is just now dying down. You step outside at the darkest time of year; clouds obscure the sky every hour till now.

The air is cold and clean. Through high bare branches of a cherry tree shine brilliant constellations. You turn slowly, recognize stars you cannot name.

The wind becomes the call you heard as a small child standing alone for one moment at the foothills of dry midnight mountains. You are bathed in the strong, silver light of a million undimmed stars.

No one knows where you've gone to, no one has yet begun to look for you. There is no measure of reassurance, no rescue coming over the long high swells of the Tehachapis.

And the child that you are, perhaps open with wounds, is more vulnerable to beauty than to ugliness. Can you see and smell and feel and hear it circling your small body in the darkness: the huge opening of grace?

O, you want this sudden loss of ledge, crave the awareness of height, welcome the path of a dream as insignificant and essential as the driest streambed. You walk into it and become so dark and full that only a glimpse of starlight glancing off black shiny hair reveals you at last to your own family. Your grandmother, relieved, reaches to button your sweater

        but you are not warmer, though the sweater is fastened, and you are not found, though your grandmother holds you with a strong hand.

                                    You are scarred by what your eyes and soul beheld. Yet you won't give it up or give it back—what happened under those mountains, in that light, with that wind.

        The explosion of stars witnessed here moves inside you, fills your limbs, heart and bones with the radiance of the universe

                creates irrevocable change, a risky inheritance. It may not ever bring you luck, or happiness. What you see changed in the world is inescapable, undimmed, raw. But you will never be afraid again.

You have gone away from a place of shelter. You have seen the face of beauty, and sustained its deepest blow. In all your life there will be no loss, no gift, no protection, greater than that.

                                                                Go on.

Baskets

"The museum's collection . . . inspired this Chihuly series of glass work."
— *Washington State Historical Museum exhibit*

When Chihuly saw you, he thought
curve, slump, weight.
He felt the smooth sweep
of glass blown into gravity.
When I see you, I open
from an empty round place
dark as stems of maidenhair fern
or the fingers of women
who twined your strength
with reeds and tule, grass and cedar bark.

Labels gleam clean as catalogued prayers.
Twana Skokomish #157 stretches
with a belly-shaped need to hold.
Klickitat #105 rises to receive
camas root and blackberries.
Yakima Sally #24 unfolds toward water,
salmon stitched with purpose.

Indians evolve like everyone else.
I understand safety pins on regalia,
plastic pony beads,
synthetic sinew.
Times change. We grow into
what comes next.

But when I see you, baskets—
locked in cabinets,
behind glass,
preserved in shadows—
I tear wide with want
for the press of my palm
against rushes, willow, redbud;
for bear grass lips frayed and soft
against my cheek. At the edge of the room
old mouths whisper *weave, braid, fill.*
I take the coiled voices of women
into the walls of this hollow vessel.

I Dreamt Your True Name

In honor of Lorraine Escobar, Inam Mec Tanotc (Rain Cloud), Esselen Nation tribal genealogist.

I dreamt your true name.
It moved on my lips,
swirled in my mouth,
stretched out on my tongue.
I swallowed it.
The name was warm and round,
filled me. Our past

evolved in my belly:
centuries of rain
gave way to ripening.
Between my legs valleys
deepened into rivers
where you bathed
in early mornings;
my spine marked the curving
coastline of rich harvest;
across soft meadows
of my breasts, you gathered
milkweed and beargrass for weaving.
Generations flowed around us

like seasons.
We did not see
evil coming in masks of
disease, murder, displacement.
We became separated

from one another.
I could not find you
even in the intimate heat
of my blood. Your bones
came back to me broken,
scattered without ceremony.
My body bore bright scars
of extinction. I slept
but I could not die.

And I dreamt your true name
in a language worn
smooth and clean
as stones in a dry riverbed.
Follow it back to me.
I want to feel your handprints
on my skin, your teeth in my hair.
I want the dark cloud of
memory to open—
release the perfect syllables
of your birth.

Burial Ground

1.
Boys from the Andes:
white cotton shirts with long sleeves
buttoned tightly at slender wrists
or rolled up over taut forearms.
Slim hips in black pants, a gleaming
braid down each back. One plays
guitar, one drums, two hold flutes
of simple design. Their voices rise
from throats not quite so young
as their faces, and the color of skin
reminds me of dreams I've had
where pleasure rides me like wind
over hills of dry grass, rippling
down each stalk into the earth.

2.
Deeper, stirring dust, wind slips
into tiny holes and passageways
made by ants, spiders, earthworms—
breathing down, breathing down
to where soil darkens, becomes wet
with delicate moisture, the distant
seepings of a river, or that subterranean
place of origins where Indian voices
come into this world. Now,
standing on a cement sidewalk
I remember that dream: hear
music so old it has aged these lovely

dark boys past all reckoning,
past beauty, or burial.

3.
I know how the earth remembers
the bones of a river—
cool hollows where trout rested
or released eggs, rocks that bear
faces once stroked by water's constancy.
The soil retains tule roots, skeletons
of moss, even as dust gathers and moisture
retreats past its own unjust grave.
Always, the voice of a river seeks
an instrument for return. I know why
tall grass covers these hills, dries brown
on the movements of an absent current.
Silence ripples subversively beneath the land
like the oldest music of all.

Waking

"the beauty of darkness/is how it lets you see"
—*Adrienne Rich*

Darkness is my sister
the elder one
who teaches me songs
and knows the constellations
flung above black
sweet-smelling mountains

darkness is my brother
the little one
who asks me to hold
his square brown hand,
so afraid of his own color,
motherless, wanting love

darkness is my homeland
my origin, my grave—
all the history I need.
When I braid my hair,
whole tribes recite genealogies
between the strands.

It is good
to know my place
and trace all the paths
over and over
to find my way
by echo, taste of riverscent

and breeze,
not relying on light
to find the bones of my ancestors:
every sturdy limb
close as my own shadow.

Here in the dark
nation of my body
I am never homeless.

Deborah A. Miranda is a mixed-blood woman of European/Esselen-Chumash descent, and a member of the Esselen Nation. Her poetry, essays and articles have appeared in *Bricolage, Calyx, Callaloo, The Cimarron Review, Raven Chronicles,* and *Sojourner,* as well as the anthologies *Bad Girls/Good Girls: Women, Sex and Power in the Nineties* (Rutgers U. Press, 1996) and *Durable Breath: Contemporary Native American Poetry* (Salmon Run Press, 1994). In her various careers, Ms. Miranda has been a student, a teacher of severely handicapped special needs children, and a housecleaner. Currently, she is working on an M.A. in English at the University of Washington. She and her husband are raising two children in Tacoma, Washington.

photo: Sally Nole

First Book Awards For Poetry

Established in 1992 in conjunction with the Returning The Gift Festival, The North American Native Authors Poetry Award is given for a first book by a Native writer. Named the Diane Decorah Award in memory of a Native writer and supporter of other Native authors, its winners published by The Greenview Review Press are:

1992   Gloria Bird *Full Moon On The Reservation*
1993   Kimberly Blaeser *Trailing You*
1994   Tiffany Midge *Outlaws, Renegades and Saints*
1995   Denise Sweet *Songs For Discharming*
1996   Charles G. Ballard *Winter Count Poems*
1997   Deborah A. Miranda *Indian Cartography*
1998   Jennifer Greene *What I Keep*